URBAN
ECOLOGICAL
Planning Guide

for Santa Clara Valley

SFEI San Francisco Estuary Institute

OPEN SPACE AUTHORITY
SANTA CLARA VALLEY

Prepared by:

San Francisco Estuary Institute

In cooperation with and funded by:

Santa Clara Valley Open Space Authority

Additional funding:

Google

Authors:

Steve Hagerty

Erica Spotswood

Katie McKnight

Robin Grossinger

Design:

Ruth Askevold (Design and Production)

Katie McKnight (Illustration)

SFEI PUBLICATION #941

SUGGESTED CITATION

Hagerty S, Spotswood E, McKnight K, Grossinger R. 2019. *Urban Ecological Planning Guide for Santa Clara Valley*. Prepared for the Santa Clara Valley Open Space Authority. SFEI Publication #941, San Francisco Estuary Institute, Richmond, CA.

REPORT AVAILABILITY

Report is available online at www.sfei.org

IMAGE PERMISSION

Permissions rights for images used in this publication have been specifically acquired for one-time use in this publication only. Further use or reproduction is prohibited without express written permission from the individual or institution credited. For permissions and reproductions inquiries, please contact the responsible source directly.

COVER IMAGE CREDITS

Cover photographs by Shira Bezalel (SFEI) and illustration by Katie McKnight (SFEI).

ACKNOWLEDGEMENTS:

SFEI is grateful for the thoughtful project direction, valuable input and thorough review by Santa Clara Valley Open Space Authority staff, including Marc Landgraf, Jake Smith, Donna Plunkett, Matt Freeman, and Andrea Mackenzie. We are thankful for additional project funding support from Google. We also are indebted to Dan Stephens and Vicki Chang from H.T. Harvey & Associates for review of the initial draft report, and Junko Bryant, Deanna Giuliano, Paul Heiple and Nikki Hanson from Grassroots Ecology and Nara Baker from Our City Forest for review of plant lists. Finally, the report benefitted from contributions by additional SFEI staff, including edits by Matt Benjamin, cartographic assistance from Micha Salomon, and additional input from Micaela Bazo, Amy Richey and Erin Beller.

CONTENTS

URBAN BIODIVERSITY • PHOTOS BY SHIRA BEZALEL (SFEI)

Introduction

Like most cities, the urbanized region of Santa Clara Valley is a challenging place for plants and animals to make a home. Largely covered with pavement, crisscrossed by major freeways, and fragmented by a variety of land uses, the urban landscape creates barriers to the movement of wildlife and hostile environments for plants. While a small set of species tolerant of cities (such as pigeons and raccoons) can tolerate these difficult conditions, our cities have the potential to support much greater biodiversity. Urban greening projects are already occurring piecemeal across urban landscapes. Harnessing this momentum can help these efforts build greater benefits for biodiversity and for people.

Urban greening projects may take many forms, such as rain gardens beside roadways; bike trails with vegetated medians; planted and expanded riparian vegetation along stream corridors; and landscaping in corporate campuses, municipal parks and private gardens. While often driven by a narrow set of goals such as stormwater capture, public safety, shade, or beautification, these green spaces can be designed and coordinated to support enhanced biodiversity as well. Rain gardens can support native wetland plants, bike trail medians can support ribbons of wildflowers for native insects, street trees can shade pedestrians and provide acorns for birds and squirrels, and stream revegetation or daylighting projects can provide a reflective place for people and strengthen corridors for regional species movement. While each of these types of projects provides benefits, strategically designing these features to reflect ecologically-minded planning, and where possible, coordinating across the landscape, can provide value for humans and nature alike.

This document provides some of the scientific foundation needed to guide planning for urban biodiversity in the Santa Clara Valley region, grounded in an understanding of landscape history, urban ecology and local setting. It can be used to envision the ecological potential for individual urban greening projects, and to guide their siting, design and implementation. It also can be used to guide coordination of projects across the landscape, with the cooperation of a group of stakeholders (such as multiple agencies, cities and counties). Users of this report may include a wide range of entities, such as local nonprofits, public agencies, city planners, and applicants to the Open Space Authority's Urban Open Space Grant Program.

This document is not intended to inform all aspects of site-specific planning. Much guidance on these topics is already available, and we reference some of those resources here. Rather, it is intended as a companion to existing materials, to inform a broader vision of how such site-scale projects can fit into the larger fabric of the Santa Clara Valley landscape. This document also recommends appropriate habitats and lists of their associated plants for the region, as well as general guidance on practical considerations related to project implementation.

In Chapter 1, "Local Setting," we explore the landscape context of Santa Clara Valley by drawing on historical ecology—the study of ecological patterns prior to extensive development—to develop a deeper understanding of landscape patterns and processes, and how they have changed over time.

In Chapter 2, "Coordinated Planning and Implementation," we outline approaches for coordinating projects to support biodiversity across the landscape. In addition, we discuss how urban greening actions and pathways to implementation differ across land uses and stakeholders, and how to integrate them on a project and programmatic level.

In Chapter 3, "Planting Considerations," we explore how to use historical and contemporary information to select habitat goals, provide plant lists to guide the building of these habitats, and consider other factors that may impact plant selection and sourcing.

Finally, in Chapter 4, "Practical Considerations," we discuss practical, planning and policy considerations that may affect on-the-ground implementation (ranging from community input, approved species, infrastructure and site management) and provide a list of supporting resources to address these. The information provided is most pertinent to Santa Clara County, but the broad approach and guidance could be useful in other geographies.

RESILIENCE

Ecological resilience is the ability of a landscape to sustain ecosystem processes and biodiversity over time, despite change and uncertainty (Beller et al. 2015, Beller et al. 2019). The **Landscape Resilience Framework,** described under "Related Resources," offers a broad overview of elements to consider for ecological-based resilience planning. Applied to our region, the concept of ecological resilience might be defined as the ability to sustain Santa Clara Valley biodiversity through future climate change and city-specific stresses (e.g., development, pollution, altered food webs and invasive species). Actions taken to support urban biodiversity, which include improving tree canopy, vegetation cover, habitat quality and connectivity of habitat patches in the city, may not only improve the resilience of urban biodiversity, but also of natural and human systems to rapid climate and urban change.

NATIVE PLANTS IN THE CITY • PHOTOS BY SHIRA BEZALEL (SFEI)

DEFINITIONS

Biodiversity: The abundance and diversity of biological life. Here, we focus primarily on native biodiversity, or the plants and animals native to a particular geography. In this document, we use the term biodiversity interchangeably with "nature" and "plants and animals/wildlife."

Urban: an area of high human population, significant level of built infrastructure, with a low proportion of intact native ecosystems relative to land area. Urban includes high density downtown cores, as well as lower density industrial or suburban areas to the fringes of cities. In this report, "urban areas" and "cities" are used interchangeably.

Urban greening: primarily refers to the practice of increasing vegetation cover in a city, which could take a variety of forms, from stormwater infrastructure to street trees. It also encompasses a variety of less direct improvements to natural infrastructure or biodiversity, such as water and energy efficiency improvements and wildlife-friendly management practices such as bird-friendly window design and reducing pesticide use.

Santa Clara Valley: The geography of interest for this document, this terms refers to the urbanized area of the valley floor. For the purposes of this report, the urban area of interest is represented by the Open Space Authority's Urban Open Space Grant Programs area, roughly corresponding to the cities of Campbell, Milpitas, Morgan Hill, San Jose, and Santa Clara, as well as their surroundings. This approximate study area can been seen on page 11.

Patch: a unit of vegetated open space that can provide habitat for plants and animals. A patch can be small, such as a downtown pocket park with a few trees, or large, such as a regional open space preserve near the edge of the city.

URBAN GREENING • PHOTO BY SHIRA BEZALEL (SFEI)

RELATED RESOURCES

This document is supported by several other related reports authored by the San Francisco Estuary Institute. Together, these resources guide how to begin to address the challenge of building ecological resilience across landscapes, while providing a scientific basis for why and how to integrate ecosystems into urban settings.

Our approach takes a landscape perspective, building on the *Landscape Resilience Framework* and the framework's regional application in the *Vision for a Resilient Silicon Valley*. We use historical ecology information and ecological guidelines from *Re-oaking Silicon Valley* and the *Urban Biodiversity Framework* to inform how to support biodiversity in urban areas. Together, these companion documents provide a deeper understanding of the historical landscape, the transformation and changes in the valley, and the scientific context that are mentioned in this report in brief.

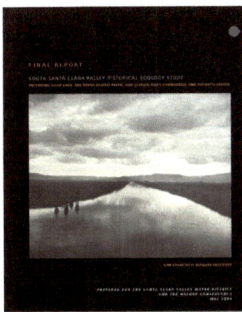

Santa Clara Valley Historical Ecology studies. This series of reports reconstructs the historical ecology of Santa Clara Valley through studies of the Coyote Creek watershed, Western Santa Clara Valley, and South Santa Clara Valley. These reports provide deep and richly illustrated research on the historical extent, distribution, and characteristics of habitats in Santa Clara Valley that can be used to understand the ecological potential of the contemporary landscape [Grossinger et al. 2006, Grossinger et al. 2008, Beller et al. 2010]. www.sfei.org/projects/santa-clara-valley-historical-ecology-gis

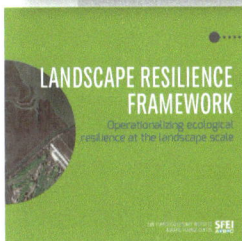

Landscape Resilience Framework. This report, and its associated scientific publication, provides a set of seven elements critical for building landscape-scale ecological resilience (the ability for a region and its nature to be robust and flexible enough to persist and evolve over the long run). The elements are intended to provide practical considerations for the management of highly modified landscapes, synthesized from the scientific literature [Beller et al. 2015, Beller et al. 2019]. resilientsv.sfei.org/content/resilience-framework

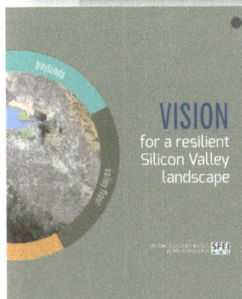

Vision for a Resilient Silicon Valley Landscape. Applying guidance developed in the Landscape Resilience Framework, this report identifies the broad components of ecosystem health and resilience for the region's baylands, uplands, creeks and urbanized valley floor, building on historical ecology studies for context [Robinson et al. 2015]. www.sfei.org/projects/resilient-silicon-valley

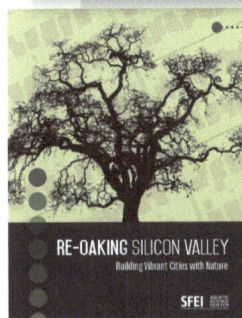

Re-Oaking Silicon Valley. This report applies principles from the Landscape Resilience Framework, Resilient Silicon Valley, and data from Santa Clara Valley historical ecology studies to offer conceptual guidelines for restoring elements of oak woodland ecosystems in urban settings to provide benefits for people and nature. The report also provides some specific design and planning recommendations for implementation, with considerations for wildlife support, genetic connectivity, species life history components, and ecosystem function [SFEI 2019]. www.sfei.org/projects/re-oaking

Urban Biodiversity Framework. The Urban Biodiversity Framework draws on urban ecology research from around the globe to identify how to plan for biodiversity support at the city-scale. The project identifies specific elements responsible for supporting biodiversity in urban settings, using Silicon Valley as an example [SFEI 2019]. www.sfei.org/projects/healthy-watersheds-resilient-baylands

1 Landscape Ecology, Change and Land Use of Santa Clara Valley

From oak groves and wildflower meadows to subdivisions and office parks, Silicon Valley has experienced dramatic changes in land cover over the past 170 years. These pages summarize some of the historical and current features of the urbanized region of the Santa Clara Valley.

The Santa Clara Valley is nestled between the Diablo Mountains to the east, the Santa Cruz Mountains to the West, the Baylands of southern San Francisco Bay to the north, and agricultural fields to the south. A dozen stream channels cross the valley floor. The area experiences a Mediterranean climate with winters that bring atmospheric rivers. These storms carry most of the season's rain and streamflow in just a few storm events (Dettinger et al. 2011) while summers are warm and dry. Yet, while dominant geographic features have remained somewhat constant, the distribution of habitat types and land uses has shifted significantly.

Historical ecology can serve as a foundation for understanding landscape patterns and processes. Future urban ecological planning for biodiversity should not necessarily aim to replicate the historical landscape, particularly given the scale and magnitude of historical transformation and predicted climate change. However, historical ecology can help understand trends, establish locally-relevant restoration targets, and identify the potential for habitat types to be restored in different areas. In combination with information on contemporary conditions and projections of future climate change, this resource can provide a framework for prioritizing and coordinating urban greening projects across the landscape.

SAN JOSE, IN THE VICINITY OF SAN JOSE MUNICIPAL GOLF COURSE, 1939 (ABOVE) • 2010 (BELOW) • COURTESY OF USDA (HISTORICAL IMAGERY ORTHORECTIFIED BY SFEI 2005)

THE LANDSCAPE THEN

Historically, the valley floor consisted primarily of well-drained alluvial soils (fine-grained deposits from streams) deposited in alluvial fans and natural levees. At lower elevations closer to the Bay (and in other natural basin areas, e.g., near Willow Glen, San Jose), soils were higher in clay content, drained more poorly, and held higher ground water tables. These gradients supported a mosaic of habitat types. Oak woodland, savanna and grassland spread out from the foothills into the valley, while intermittent and perennial streams supported corridors of riparian vegetation upstream before spreading out into wet meadows in lower elevation areas, or flowing into the Baylands. Alkali meadows lay on the fringes of the Baylands, and in some cases, along or near intermittent water bodies (see Figure 1.1). Local Native American tribes contributed to local habitat diversity by maintaining open landscapes using periodic controlled burns (Beller et al. 2010, Grossinger et al. 2008). Such a rich mosaic of vegetation and hydrology supported a wealth of flora and fauna (Grossinger et al. 2006, Grossinger et al. 2008, Beller et al. 2010).

Much has changed over time. While Muwekma Ohlone and other native peoples continue to reside in the area, they have lost much of their legal access and rights to practice traditional land management techniques (such as prescribed burning) that shaped much of the landscape prior to colonization (Anderson 2005). Over one million people now inhabit Santa Clara Valley. Soils and waters have been altered dramatically in many areas. Artificial fill lays atop clay soils near the Bay, and large areas of the valley floor have been paved over, graded, compacted, and contaminated. Groundwater levels have dropped in some areas and rebounded in others. As a result of these disturbances, elevations have subsided in some areas. However, current groundwater recharge efforts and water imports have stabilized elevations regionally (Galloway et al. 1999). A vast expanse of habitat has been lost, fragmented, and converted to other land uses. On the valley floor, suburban, low-to-medium density development now occupies the majority of the land area. Commercial and industrial areas are concentrated along major roads and highways at lower elevations near the Bay. The surrounding foothills support a patchwork of agriculture and protected open space. Several major streams have been straightened and disconnected from the Baylands, with their waters now running perennially. The watersheds and hydrology of virtually all the valley's creeks and rivers have been dramatically altered by artificial drainage, paving, dams, and other hydromodifications. Clearing, paving, and invasive species have reduced the cover of native plants in many areas. Roads have fragmented the remaining habitat patches, while noise and light impact animal presence and behavior in and around cities (Diamond and Snyder 2016, Forman 2016).

Figure 1.1: Historical ecology map. Historical ecology is the study of nature over with time, often focusing on reconstructing the types of habitat and natural processes that occurred at one point in time on the landscape. This map represents the distribution of historical habitat types circa 1850, based on the historical ecology studies described on page 5. The generalized highlighted study area is based on Santa Clara Valley Open Space Authority's Urban Open Space Grant Program eligible grant area.

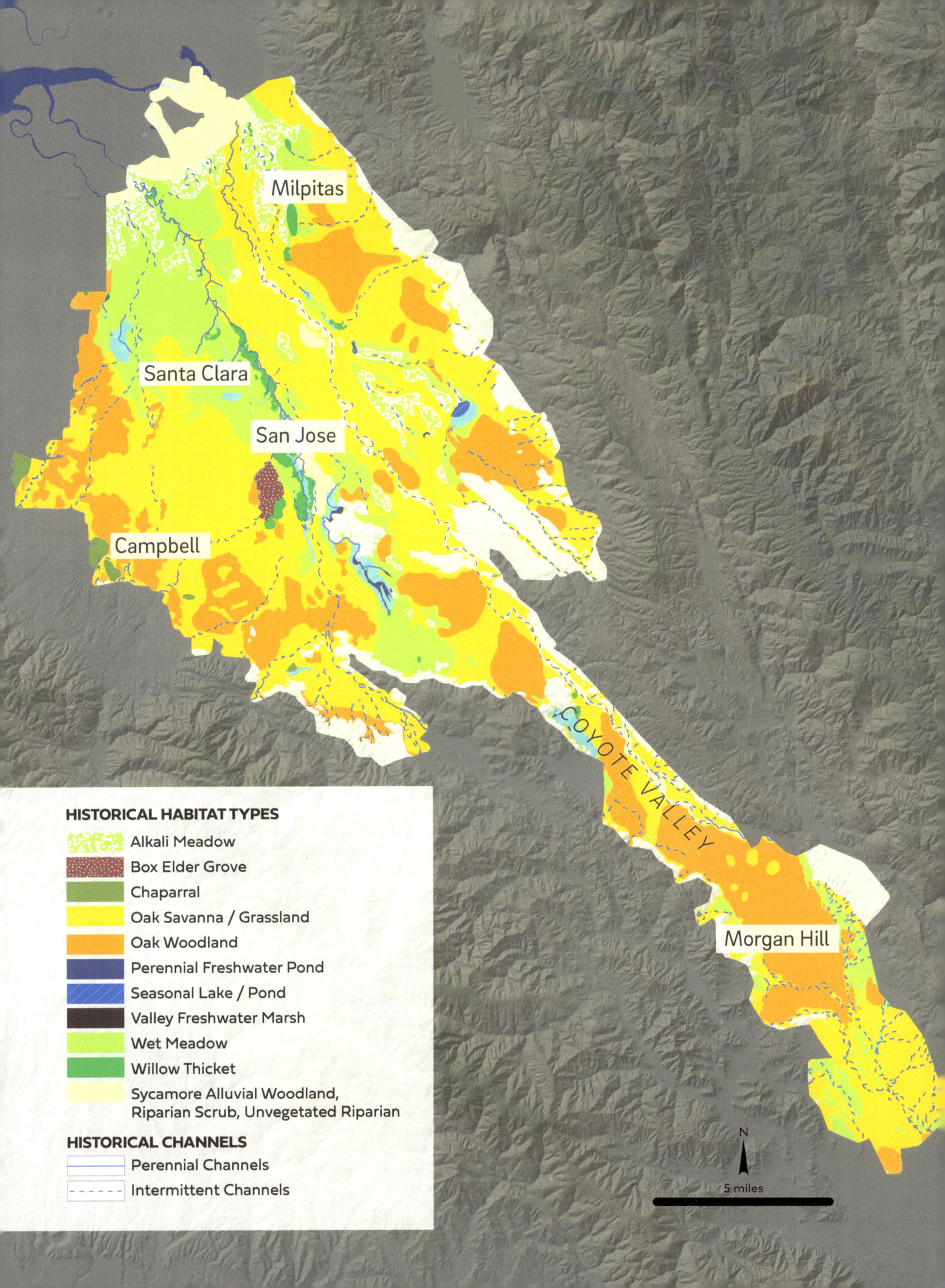

Milpitas

Santa Clara

San Jose

Campbell

COYOTE VALLEY

Morgan Hill

HISTORICAL HABITAT TYPES

Alkali Meadow

Box Elder Grove

Chaparral

Oak Savanna / Grassland

Oak Woodland

Perennial Freshwater Pond

Seasonal Lake / Pond

Valley Freshwater Marsh

Wet Meadow

Willow Thicket

Sycamore Alluvial Woodland,
Riparian Scrub, Unvegetated Riparian

HISTORICAL CHANNELS

Perennial Channels

Intermittent Channels

N

5 miles

THE LANDSCAPE TODAY

Nevertheless, the valley still provides habitat for an impressive variety of native species—hundreds of species of mammals, birds, reptiles, amphibians, freshwater fish, invertebrates and plants (County of Santa Clara et al. 2012). The large swaths of protected Baylands, open space, and agricultural lands provide habitats for species that require large territories or are sensitive to human activity, such as the mountain lion and Ridgway's rail. Stretching from the hills to the Baylands through the city, riparian areas provide fragmented corridors for foraging and movement for many songbirds and mammals small and large (e.g., coyotes). Larger patches along riparian corridors provide refuge for some ground-nesting birds (e.g., California Quail). Even in the urban core, backyards, schoolyards, urban parkland, and utility corridors continue to support small animals such as the monarch butterfly, arboreal salamander, and many migratory bird species. Opportunity is ripe for investing in urban greening projects that bring more nature into cities and make it is more accessible to people.

The contemporary landscape of Santa Clara County encompasses a variety of land uses, including residential, office park, urban open space, and transportation corridors. Development has not been entirely random. In fact, these patterns generally reflect historical landscape ecological patterns. Somewhat open, well-drained areas such as former oak woodland, savanna, and grassland were often transformed first to orchards, and then to subdivisions. Large mosaics of bayside wetlands in low-lying areas of high groundwater were often first cropped as hay fields and later filled to support industrial and commercial development (Grossinger et al. 2006, Grossinger et al. 2008, Beller et al. 2010). Other forms of urban development follow ecological patterns less strictly but are nonetheless predictable, such as commercial development clustered along transportation corridors.

All of these land use types have opportunities for urban greening that may benefit biodiversity, though they may vary by jurisdiction and local zoning. Section 2 will demonstrate in more detail what types of opportunities are available in different land uses, and how to coordinate multiple urban greening actions across the landscape to strategically benefit biodiversity as well as to logistically implement multiple adjacent projects.

Figure 1.2: Land use in the Santa Clara Valley. This map represents the distribution of land use types across the valley floor, based on data from the Association of Bay Area Governments. Most urban land is dedicated to low-to-medium density residential areas, with commercial districts distributed along key transportation corridors, and industrial areas and office parks near the Bayshore. The generalized highlighted study area is based on Santa Clara Valley Open Space Authority's Urban Open Space Grant Program eligible grant area.

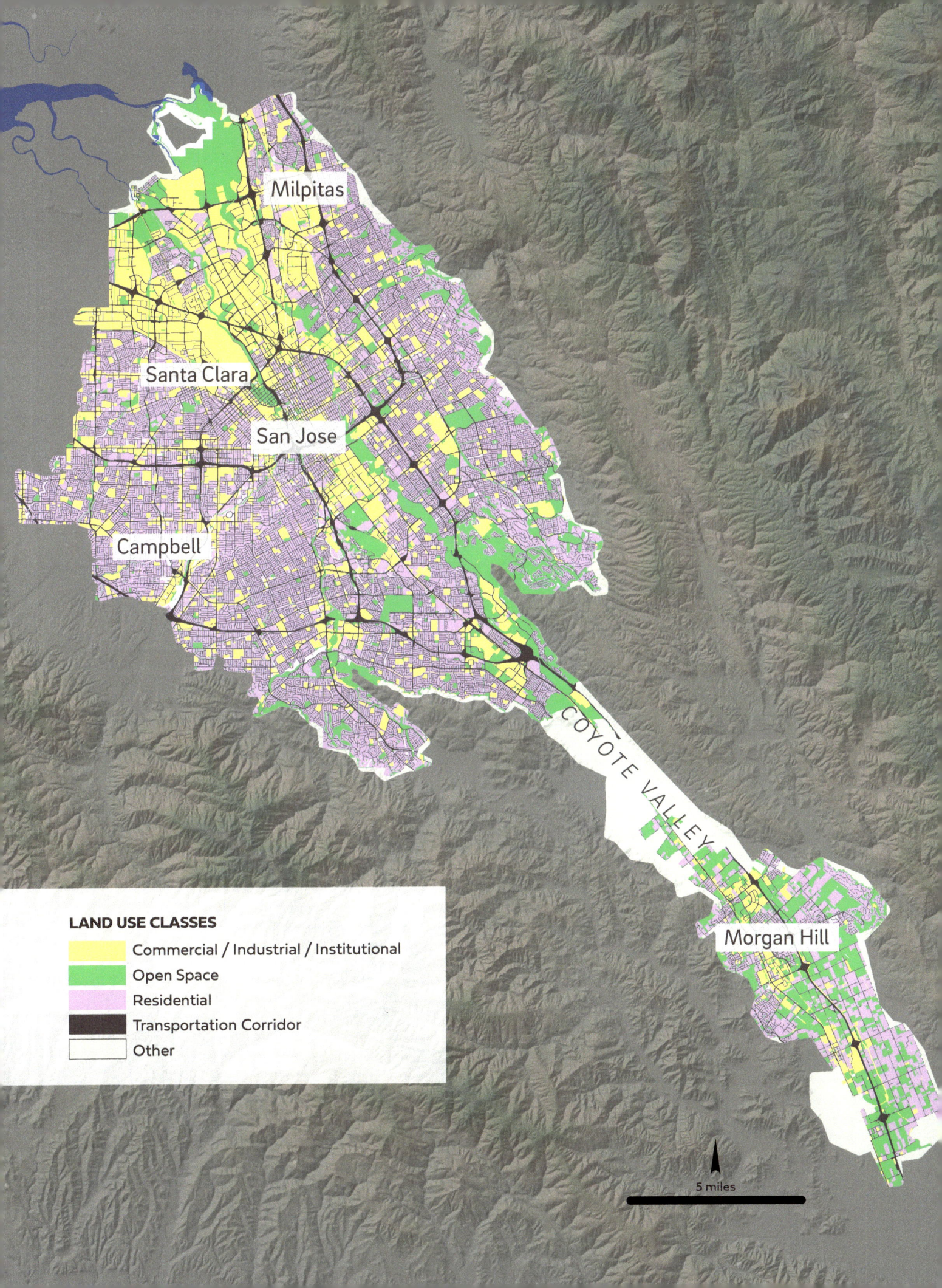

LAND USE CLASSES

- Commercial / Industrial / Institutional
- Open Space
- Residential
- Transportation Corridor
- Other

Milpitas

Santa Clara

San Jose

Campbell

COYOTE VALLEY

Morgan Hill

5 miles

2 Coordinated Planning and Implementation

Urban greening projects in cities are often implemented independently and not directly coordinated. While large patches of protected open space often offer the best homes for plants and wildlife, there is opportunity to scale up even small ecological improvements in a coordinated way that provides broader benefits to biodiversity. For instance, Goddard et al. (2010) suggest that:

> "If...green spaces can be spatially arranged to maximise total habitat patch area and minimise isolation, this will result in benefits to urban biodiversity... It is therefore imperative that gardens are not viewed as separate entities...but instead managed collectively"

In other words, implementing and managing multiple urban greening projects at different sites can work to increase the support of biodiversity across the broader landscape. Birds that forage in a native plant garden in a neighborhood park can build their nests in native trees planted by an urban forestry program in a neighboring area. Coordinated greening actions can help overcome the extreme ecological fragmentation of urban landscapes.

Given the diversity of land uses, jurisdictions, and community interests present in Santa Clara Valley, implementing urban greening actions can be complex. For example, managing native trees on streets and in parks might require collaboration among city forestry and public works departments to maintain trees in the public right of way, and with the parks department to do so in open spaces. A streetscape master plan may guide what can be planted in which locations, and community groups may inform the best use of the park. Analogous interactions can play out on a broader scale, such as if multiple cities coordinate their master plans for parks along a creek that crosses their boundaries. The following sections discuss how Santa Clara Valley might approach regional and local planning across land use types and stakeholder engagement to achieve biodiversity and ecological resilience goals.

Figure 2.1 demonstrates an example distribution of urban greening projects across a neighborhood that could support biodiversity, and describes individual urban greening actions in detail. "Planning for biodiversity" in Chapter 2 and Figure 2.2 highlight how a subset of these actions can be planned or work together to support biodiversity goals. Finally, "Planning across stakeholders and by land use type" explores how to implement these actions in each land use as well as structurally.

URBAN GREENING • PHOTO BY SHIRA BEZALEL (SFEI)

Green roofs

When installed with native plants, green roofs can provide habitat for flying insects and birds. Green roofs can also capture rainfall and mitigate runoff, helping to improve water quality and reduce flood risks.

Vegetated bike paths and medians

Bike paths can connect parks to one another, linking ecological features. Along roads, separated bike lanes can be planted with native trees, shrubs, and flowers, and can incorporate bioswales and other green infrastructure. These elements can provide shade, reduce air pollution, and protect road surfaces while improving safety by helping to separate bikes from cars.

Residential gardens

Native trees, shrubs, flowers, and other habitat features such as bird boxes and water fountains can all be incorporated into residential yards. These enhancements can help build local habitat complexes that can support ecological communities. Large trees can help reduce cooling costs by shading houses and reducing runoff. Gardens provide beauty and respite for people, and can reduce water use if lawns are converted to drought-tolerant native plants.

Rain gardens and other green infrastructure

Rain gardens and other green infrastructure can be integrated along small and medium sized roads and in parking lots. These features can be planted with native plants to increase their ecological value. Bioswales capture urban runoff, improving water quality and helping to mitigate flood risk.

Parking lot, street, and traffic circle trees

Native trees in parking lots and traffic circles can provide forage and refuge for passing birds, insects, and some small mammals between patches of habitat. Parking lot trees provide much-needed shade in the summer, and along streets and in traffic circles, trees can calm traffic and shade streets. Trees also capture and store carbon, capturing some of the emissions responsible for global climate change.

Landscaping in commercial spaces

Large patches of habitat can be integrated in commercial landscaping, such as through native trees and bioswales in parking lots. These features can support wildlife while providing shade and beauty, capturing carbon, and reducing runoff.

Habitat restoration in public parks

Public parks can act as major hubs for biodiversity in the urban landscape, particularly if they contain large patches of high quality habitat. These patches can support unique wildlife not found elsewhere in the urban landscape. Parks also provide residents with access to large green space, important for both mental well-being and physical health.

URBAN GREENING
PHOTOS BY SFEI

Corporate Campus
[Research, Office Parks &
Light Industrial]

**Single Family
Homes**
[Medium Density
Residential]

Commercial

**Multi-family
residential
(apartments)**

**Single
Family Homes**
[High Density
Residential]

Shopping Center
[Commercial]

Commercial

Park
[Open Space]

**green
roof**

Figure 2.1: Urban greening opportunities. Conceptual diagram of opportunities for urban greening projects across the landscape. Scale is 1 : 1000.

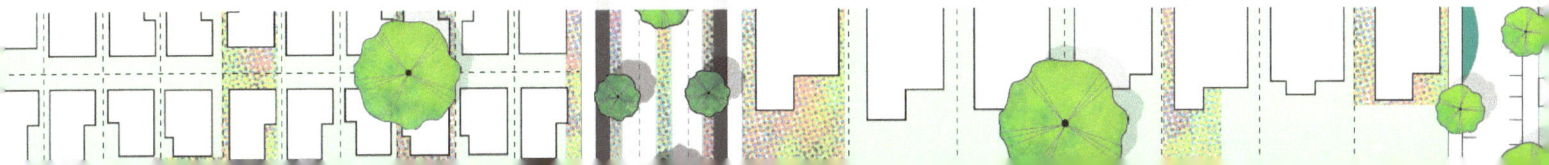

PLANNING FOR BIODIVERSITY

To build urban landscapes that sustain biodiversity at scale, multiple elements should be considered. Scientific literature suggests that creating and maintaining large urban green space patches, creating and improving large regional corridors between patches, and weaving habitat into the fabric of the city (through multiple urban greening actions) can all work together to support biodiversity in cities (Beninde et al. 2015). Improving habitat quality by establishing native plants, creating complex vegetation structure, and promoting wildlife-friendly management practices (e.g., retaining dead organic matter and reducing pesticide use) can also increase the biodiversity value of existing and planned urban greening projects (Aronson et al. 2017, Goddard et al. 2010, Threlfall et al. 2017).

Some of these elements help certain groups of species in particular. For instance, large patches can benefit species sensitive to urbanization or with larger home ranges, such as bobcats. Management practices such as retaining leaf litter on the ground can benefit soil invertebrates and ground-foraging birds.

Coordinated conservation projects can provide more value than the individual actions they involve. For instance, tree planting projects arranged in a grove across shopping centers and streets can support a colony of acorn woodpeckers. Oaks placed closely together, across different properties, can allow butterflies to travel from tree to tree. Planting linear strips of native wildflowers along streets and yards can work in concert to provide habitat and connectivity for pollinators. Creating habitat in backyards surrounding a small park can increase the value of the park itself. Thus, the integration of these actions, within patches and across an urban area, is necessary to support a diversity of species in the city. Figure 2.2 demonstrates how this type of coordination in a neighborhood can create elements of a functional oak woodland to support a wide variety of species. For further exploration of these concepts please see Re-Oaking Silicon Valley and the Urban Biodiversity Framework (Spotswood et al. 2017, SFEI-ASC 2019).

Acorn woodpeckers can tolerate urban landscapes, including San Jose, if native oak trees are present. Providing a *large tree* (32"+), wherein woodpeckers can store acorns, amidst a grove of approximately *20 oak trees scattered across 15-20* acres can support a new acorn woodpecker colony (Spotswood et al. 2017). The grove of smaller oaks can span a variety of land uses, including front yards, shopping centers, and streetsides.

Arboreal salamanders often use large oak trees as refuges to lay their eggs and estivate. Downed logs and leaf litter in urban parks can provide habitat for the salamander to forage and rest. Arboreal salamanders tend to use a home range smaller than an acre (Morey 2014). Considering needs of many small home range creatures, *parks of 2 acres or larger* can play a major role in supporting biodiversity and tend to house significant numbers of species (SFEI-ASC 2019, eBird 2019). In addition, the value of parks can be increased if their surroundings also provide some habitat value. *Improving habitat surrounding park habitat patches*, such as in backyards and along streets near parks, via planting oak trees and maintaining leaf litter and downed logs, can lead to greater cumulative benefits than a park or yard might provide on its own.

The **California hairstreak** is now rare in the region, but the urban landscape can play a role in its recovery. California hairstreaks lay breed and lay their eggs within native oak trees such as the valley oak. As caterpillars, the hairstreak consume leaves from oaks and other trees. The species requires oaks in close proximity to one another to disperse across the landscape. *Spacing native oaks 500 feet or less apart,* particularly in linear corridors, can support movement of the hairstreak while also enhancing pollination rates among trees (Spotswood et al. 2017, A. Shapiro pers. comm.).

Monarch butterflies and other butterfly species rely on nectar plants distributed across the landscape for sustenance as they travel. Native milkweeds (e.g., *Asclepias fascicularis*) are particularly important for Monarch caterpillars, which feed exclusively on the plant genus. Planting *linear corridors of native wildflowers* can potentially support butterflies such as Monarchs, which use continuous habitats (Kasten et al. 2016, A. Shapiro pers. comm.). Linear corridors can run alongside residential streets (e.g., in yards, medians, or street planters) and can connect to larger parks with butterfly habitat. However, wider, more protected corridors are likely necessary to support the movement of larger or more urban-sensitive wildlife.

Figure 2.2. Example coordination of urban greening opportunities to support biodiversity at the neighborhood scale.

To plan for biodiversity at the city-scale, across many projects, planners can use data to identify existing resources and opportunities. Such data can illuminate the distributions of biological resources (e.g., using biological surveys or citizen science data, such as eBird and iNaturalist), large patches of habitat and regional corridors (e.g., using riparian vegetation surveys, maps of stream networks, and maps of protected areas), special resources such as ponds and large trees, canopy cover, and native vegetation (e.g., using vegetation surveys, street tree inventory data, and remote sensing data).

Where feasible, planners can focus conservation efforts on locations that most effectively build upon existing ecological resources. For example, locations where open spaces can be expanded, where canopy or native vegetation cover is low, or where gaps in riparian corridors exist can all be high-priority sites for improvement.

Within a selected patch or project area, similar prioritization strategies may apply. For instance, creating dedicated patches of habitat within a larger park, and connecting these patches where possible can help these green spaces support more biodiversity. Planting native plants and providing large trees and bodies of water can provide foraging and nesting resources for a variety of wildlife. Maintaining leaves and dead branches on the ground, and reducing use of rodenticides and herbicides, can benefit the local food web. While this report focuses on coordinated design for biodiversity at a high level, more comprehensive details on how to plan spatially for biodiversity at a large scale can be found in SFEI-ASC 2019.

Actions to support biodiversity can likewise bring benefits to people. Improving tree canopy cover can capture pollutants from the air, buffer noise, and provide shade to reduce urban heat. Rain gardens and bioswales can reduce flooding and improve water quality. Small parks and gardens can provide spaces for viewing nature, which has been linked to mental and physical health (e.g, Pearson and Craig 2014, Payne et al. 1998). Coordinated urban greening, designed to best support native species, may also most benefit people. Studies link human well-being to exposure to both habitat that supports biodiversity (e.g., with mature trees and a multi-layered canopy) and exposure to biodiversity itself (e.g., Sandifer et al. 2015, Cooper-Marcus and Barnes 1995). Incorporating nature into the fabric of the city can thus help improve the quality of life within urban communities.

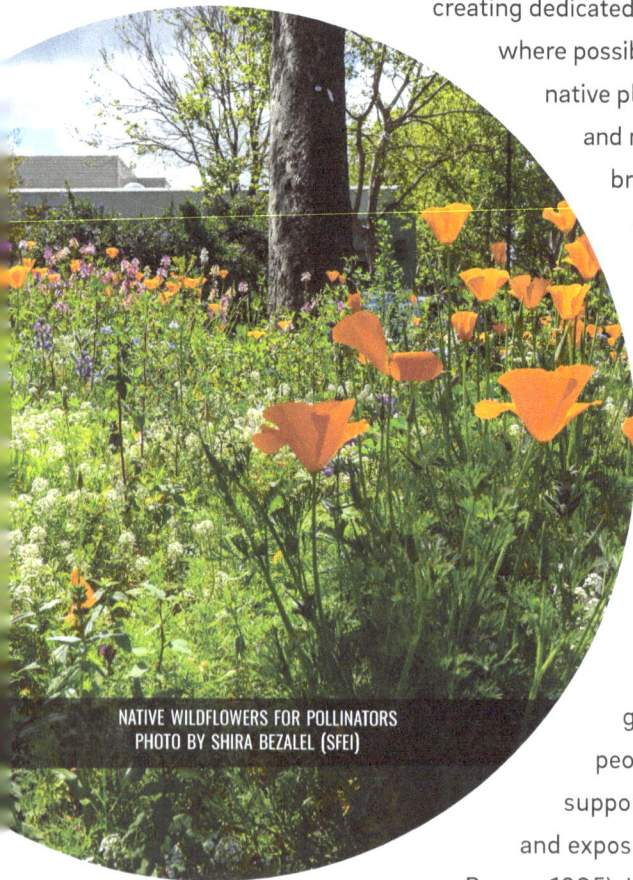

NATIVE WILDFLOWERS FOR POLLINATORS
PHOTO BY SHIRA BEZALEL (SFEI)

PLANNING ACROSS STAKEHOLDERS AND LAND USES

Coordinated improvements to support urban biodiversity across stakeholders can be achieved in a variety of ways, including local, regional and state planning processes; land acquisition; permitting; policy and regulation; education, outreach and incentive programs. At a broad level, climate action and redevelopment plans (e.g., Karlinsky et al. 2017, Romanow et al. 2018) can and do prioritize dense development and limit sprawl on the urban fringe into parks and open spaces. Local or regional regulations or policies, such as municipal or county riparian setbacks, parking requirements and developer requirements (e.g., contribution to open space), can facilitate opportunities for creation of additional urban habitat. Parks departments (such as San Jose's Parks, Recreation and Neighborhood Services), flood control agencies (e.g., Santa Clara Valley Water District) and land trusts (such as Peninsula Open Space Trust [POST]) are already working together to acquire, protect, and manage valuable parkland and waterways in the urban interior and fringe. Green infrastructure plans, policies, and permits can incorporate native species in their recommended planting lists. Urban forestry policies and streetscape master plans can prioritize native trees on recommended planting lists. Nonprofits and nurseries associated with urban forestry and other urban greening programs can prioritize supplying the region with native plants, and provide educational programming about their value, as well as incentives for planting.

Ensuring that these actions are working towards a common goal can benefit from the leadership of a coordinating entity, such as a central planning office. The City of San Francisco, for instance, has a director and program dedicated to promoting urban biodiversity across the city, including across agencies. Scientific advisors and the City Planning Office can help to set these goals for urban biodiversity. Community input is crucial to ensure that an urban biodiversity plan meets community needs and incorporates local knowledge while it benefits nature. Coordinating entities may thus benefit from interfacing with a variety of stakeholder groups as they craft biodiversity plans.

As in many cities, metropolitan Santa Clara Valley encompasses a mosaic of land uses with a correspondingly diverse array of stakeholders. Thus, coordinating multiple projects across the region may pose a significant challenge. Here, we focus on how various actors might coordinate their actions across land uses and jurisdictions to promote urban biodiversity at a neighborhood scale through individual projects, guided by local or regional planning. The following pages demonstrate how different types of urban greening actions can be incorporated within different land use types and highlighting some of the complexities of implementing each action in a given location.

Major land use types discussed here come from Association of Bay Area Governments (ABAG) land use dataset classifications, and include Research, Office Parks, and Light Industrial; Commercial; Transportation Corridors; Open Space; and Low-, Medium- and High-Density Residential.

Research, office parks and light industrial — e.g., corporate campus

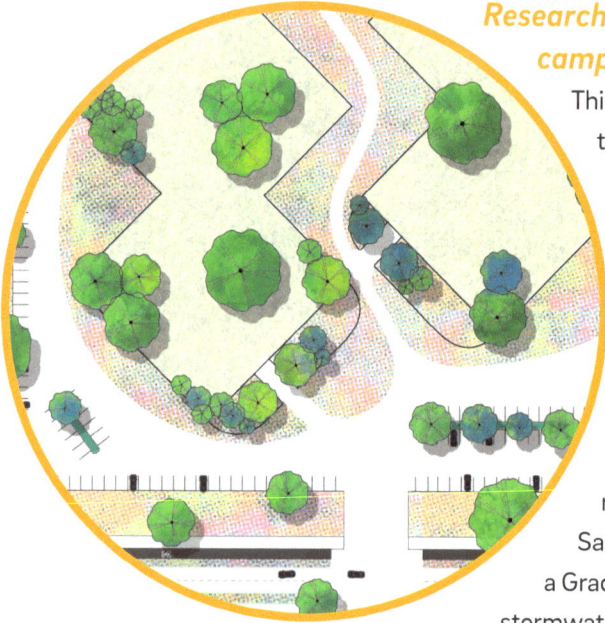

This land-use type includes extensive, open lots with office buildings near the baylands. These parking lots provide opportunities to convert asphalt to **native landscaping** and trees. The large, flat roofs of these buildings provide opportunities for **green roofs,** and the expansive parking lots give room for **large green infrastructure** such as retention basins. Large glass window panes near the baylands highlight the importance of **bird-friendly window design**. Co-benefits arising from these types of projects include additional shade for parking with more trees, stormwater capture with green roofs, and energy efficiency and associated cost savings. Implementing these actions would potentially require a permit for re-roofing to create a green roof (from the City of San Jose Planning, Building and Code Enforcement Department [PBCE]), a Grading and Drainage permit from the San Jose Dept. of Public Works for stormwater projects, and consideration of local parking minimums and working with landscape architects on plant lists to install native plants.

To incentivize these actions, cities could integrate bird friendly window building design into **building codes** and guidelines (recently completed in San Jose [Riparian Corridor Protection and Bird-Safe Design, Policy Number 6-34, City of San Jose California Council Policy 2016]). Local or state **tax incentives** for green roofs could also be improved. Finally, adjustments or compliance with additional policies such as riparian setbacks or parking minimums can increase the amount of open space to support biodiversity benefits.

Commercial —e.g., shopping center

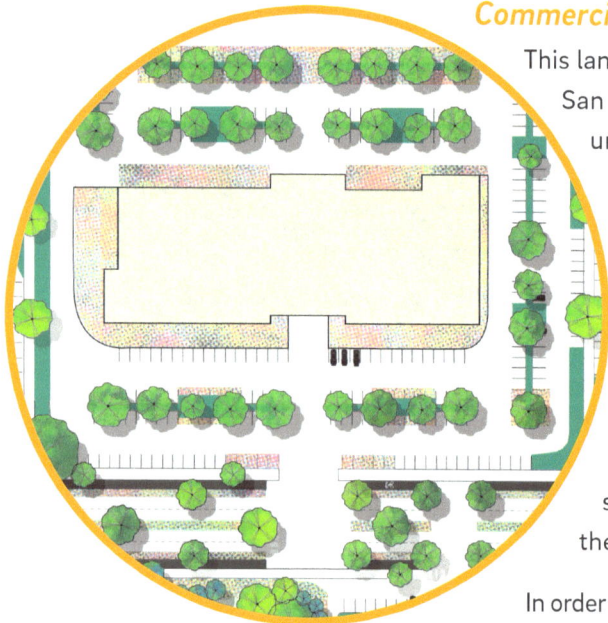

This land use type covers retail areas, including strip malls in suburban San Jose and the bottom floors of multi-use buildings in the urban core. Here, opportunities are similar to those in research, office parks, and light industrial areas: transitioning asphalt and underutilized lawn to **native landscaping** and trees, and opportunities for **green roofs**, particularly in large malls. **Small green infrastructure** such as rain gardens or bioswales with native plants can also be added adjacent to buildings or the lowest part of a parking lot. Co-benefits include additional shade for shopper parking (and potentially increased time shopping and local economic benefit as a result), as well as energy savings and stormwater capture. Implementation considerations are similar to the previous land use type.

In order to encourage these actions, **tax incentives** for green roofs could be improved, and other policies, such riparian setbacks or parking minimums can increase potential area for nature.

Transportation corridors — e.g., bike lanes and bike paths, streets, and traffic circles

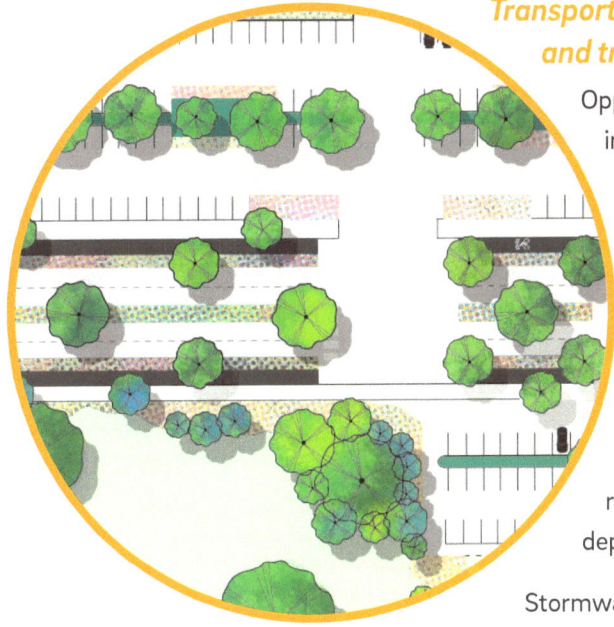

Opportunities along transportation corridors include street retrofits to incorporate **green infrastructure** (such as rain gardens or bioswales), planting native **street trees**, and **native landscaping** along streets and in medians. Converting lanes and adding additional vegetation along streets can reduce traffic speeds and improve safety, while more green infrastructure can help filter and drain stormwater runoff. Using innovative streetscape design to enhance connectivity for wildlife can also be beneficial, particularly along greenways where road crossings reduce connectivity. Individual green infrastructure projects may require permits from Public Works (cited above). Permits are also required from DOT or PBCE for planting, pruning or removing trees, depending on the jurisdiction.

Stormwater guidance documents, including municipal **green infrastructure plans**, can be one mechanism for encouraging green infrastructure that can help build biodiversity support. Specific planning frameworks, such as DOT's Better Bikeways program, can incorporate **planning and design guidance for bike lanes** that encourages installing protective buffers with native vegetation. Municipal tree manuals, such as the San Jose Tree Policy Manual, already use approved tree lists to guide appropriate trees, and are one pathway to motivate the planting of native trees. **Streetscape master plan** guidance could also include native species in their planned/desired palettes.

Open space — e.g., park

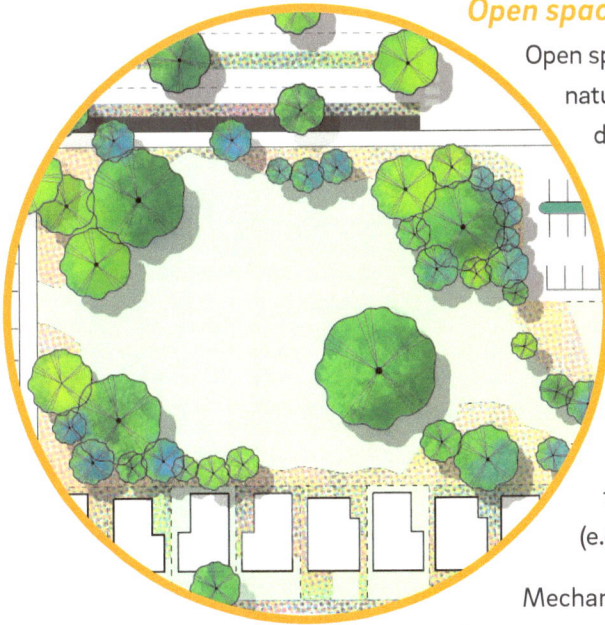

Open spaces take a variety of forms in the area, including parks, gardens, nature preserves, farms or orchards, sports fields, and golf courses. The low density of impervious surfaces allows for significant potential for habitat improvement. **Land acquisition** is a key potential action across the urban-to-rural gradient. For existing spaces, given their flexibility, **lawn removal** (especially underutilized areas at a park's edge or between its features) and replacement with native vegetation and h**abitat restoration** are key urban greening actions. Given large land blocks and significant potential for invasive species, reduction in pesticide use is another opportunity. These actions collectively can provide beauty, shade, and chances to spot wildlife. On a project level, permits may need to be coordinated for grading, tree removal, or working in or near streams (e.g., from CDFW, SFBRWQCB, USACE, USFWS/NMFS, or others).

Mechanisms to facilitate open space improvements include **zoning** new areas for open space, **pooling financial resources** among agencies, land trust, and other NGOs to acquire new open spaces. Several local organizations are already involved in this work, including POST, the San Jose Planning Commission, and San Jose City Council. Individual **parks master plans** can incorporate plans to create and protect large and connected habitat patches and native landscaping. **Grants** such as the Alliance Grants Program can help support integrated pest management (e.g., Guadalupe River Parks and Gardens, which used its award to go pesticide-free).

Medium density residential —e.g., single family homes

This land use is characterized by medium to large residences with back and often front yards. Perhaps the most significant difference homeowners can make given this is **lawn removal** and creating **native plant gardens**. Driveway and patio footprints can also be converted to vegetation. In addition, native **street trees** can be planted in the public right of way, though maintenance is typically the responsibility of the property owner. These actions can improve beauty and create opportunities for wildlife viewing, shade, and energy savings. Grading permits may be needed with any significant land alteration and permits may be required for adding, removing or pruning trees.

Local incentive programs that can motivate these actions include the Lawn Busters program of Our City Forest to **incentivize lawn removal** and replacement. Many local nonprofits are strongly connected to local communities through outreach activities, and these activities can include **education** about the value of native plants, and can provide **nursery support** to grow native plants.

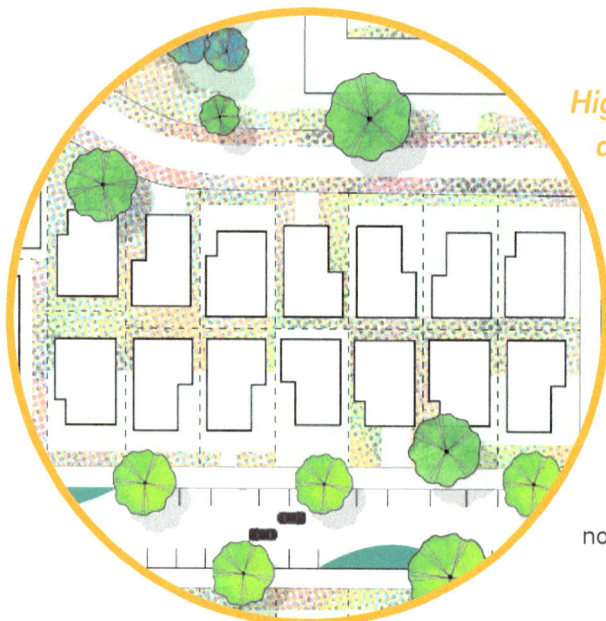

High density residential —e.g., multi-family homes and apartment buildings

This land use is characterized predominantly by multi-family homes and apartment buildings, with larger impervious surface footprints and smaller yards. **Native landscaping** is an option in small pockets where asphalt or extra parking is not required, as are native **street trees**. Permits may be required for adding, removing or pruning trees.

Opportunities for aligning biodiversity goals with existing programs are similar to those in medium-density residential, and include nonprofit outreach, education and nursery support.

3 Planting Considerations

For a given project, determining which plants to place where is not always a straightforward choice. The following section can help provide some insight on this topic. It discusses how to integrate historical and contemporary information and how to evaluate a site to set a habitat type goal. This chapter also provides a set of plant lists appropriate for several of the dominant local habitats, and provides guidance for tailoring these plant lists to a particular site, given biological, ecological and climate change considerations.

BRIDGING PAST AND PRESENT

Understanding the historical distribution of habitats on a landscape (historical ecology) can provide insight into how to design cities to support native plants and animals. For instance, certain habitat types may have been lost, greatly reduced in extent, or are now rare, resulting in declines of the species they once supported. Understanding these types of trends can help inform ecological restoration and urban greening goals.

Historical ecology can offer insight into which habitat type(s) may be appropriate for a site. The historical distribution of soil types, stream networks, and plant communities gives clues as to what types of habitat might be appropriate to recreate where—this can be one of the first resources to consult during project design. Topography, soil types and groundwater levels often remain constant in a landscape despite urbanization. Historically dominant habitat types in Santa Clara Valley included wet meadow, alkali meadow, oak woodland/savanna, grassland, sycamore alluvial woodland and sparsely vegetated riparian areas (see Grossinger et al. 2006, Grossinger et al. 2008, Beller et al. 2010). The loss of these habitats is largely responsible for losses of native biodiversity, so recovering these resources is of high value for promoting biodiversity.

While historical ecology guide what habitat type is appropriate to restore in a given site, other site-specific information is also necessary. The Santa Clara Valley landscape has radically transformed in the last 170 years, and changes to local site conditions can modify what vegetation is appropriate for a site. Historical soils, streams, and plants in the project area may have changed for a variety of reasons. Understanding how these components have changed can inform whether to restore a historical or novel habitat in a given location. The goal of urban greening under this lens is not to recreate the exact conditions of the past, but rather to integrate historical and contemporary information to design sites that are likely to support biodiversity into the future.

EVALUATING HABITAT TYPE POTENTIAL

The following section describes a set of considerations to help evaluate a site. These considerations can be helpful in determining whether historical habitat types may still be appropriate for a site, or whether land managers will have better success installing an alternate habitat type.

Existing vegetation. Observing existing plant species and their health on site or nearby can help inform what types of plants might be most suited to the site. For example, an area that was once chaparral may now support a residential street lined with sidewalk trees, including a handful of healthy oaks and other species. As elevation, soil, and groundwater may have changed here, and to take advantage of existing oak resources, adding oak woodland plants might be preferable to bringing back the shrubs of the chaparral. Similarly, another area that was once a wet meadow now may contain dense urban development, a channelized storm drain with perennial flows, and native riparian trees sprouting on the upper banks of the channel. In this case, because of the altered hydrology, as well as to capitalize on the existing tree resources, woody riparian habitat may now be more appropriate than wet meadow vegetation.

Local soil conditions. Soil type can dramatically impact what types of plants are suitable in a given plot of land. Many aspects of soils, including topsoil thickness, compaction, condition, and composition have changed dramatically in some areas from historical conditions. Identifying the cause of these alterations is important for resolving them. Common alterations include soil compaction from overuse and development, reduced soil volume and organic matter from grading, altered soil salinity from use of recycled water, increased surface irrigation, and legacy chemical or heavy metal contamination from brownfield sites. Signage, light fencing, and education can help reduce compaction from pedestrian traffic, while aeration or other decompaction methods can help to restore compacted soils in preparation for planting. Additional soil or targeted amendments (e.g., organic compost) can help improve soil health. If soils are contaminated to the extent that on-site remediation is not possible, it may be necessary to cap or excavate and replace soils entirely. Not all atypical soil conditions need to be amended, however. For instance, highly alkaline soils, especially if historically alkaline, provide opportunities to support unique alkali meadow species, without requiring significant soil modification. Soil testing can be helpful in identifying current conditions and determining the suitability of a given planting list.

Hydrology and groundwater. Some plants prefer wetter soils, while others prefer drier conditions; matching plant water requirements to site conditions is important for establishing successful restoration projects. While historical ecology provides suggested habitat types in a given area, water availability can be highly locally variable and has been altered in many areas. Groundwater pumping has increased the depth to groundwater in some places, whereas in others, local fill for development has raised site elevations, increasing the depth to groundwater onsite. Additional hydrological changes, such as storm drainages, permeability of pavement, sprinklers, and leaky pipes must also be considered. To estimate current groundwater levels, local wells can provide a proxy. If no local wells exist, the presence and condition of existing trees with deeper roots can provide additional clues as to an appropriate planting palette. If historical ecology mapping suggests wetter habitat types (such as wet meadow) are locally appropriate, but current groundwater levels are too low for this habitat type in a given area, then more terrestrial upland habitat types such as grassland and oak woodland should be considered. Poor drainage can be inferred from soil type or observed directly from standing water. In areas with historically poor drainage, consider planting more hydrophilic plants and plant communities (ex. willows). If poor drainage is instead due to land use alteration or compaction, see the previous section on local soil conditions and improving soil health. Areas that are currently maintained with frequent surface irrigation (such as lawns), but were historically disconnected from surface waters, may no longer require significant irrigation to maintain an appropriate native plant community.

PLANT LISTS BY HABITAT TYPE

The following lists provide general guidance for incorporating native plants in urban areas in the Santa Clara Valley. These lists comprise plants generally considered native to the Santa Clara Valley. They include examples of native plants related to each type of habitat historically represented on the valley floor (corresponding with the map on pg. 9) and potentially suitable to an urban setting. Historical habitat types provide baseline information that can be reconciled with contemporary information to evaluate the appropriate habitat type for a site. Once a habitat type has been selected, the tables in the following pages can be used to find potentially suitable plants.

The lists include recommendations for appropriate understory plants, shrubs, and trees. These lists are meant to serve as inspiration and represent species that are both widely available and likely to succeed in urban landscapes. Plants often tolerate a variety of conditions, and do not always neatly segregate into vegetative communities—therefore, some species are present on multiple lists. While the lists provided represent a diversity of plants in terms of taxonomy, structure, and bloom times, they are not exhaustive.

Native California vegetation is highly diverse, and other resources list many additional suitable species (see Useful Resources section, pages 35-36). Furthermore, some non-native species have value for wildlife, but these are not listed here, as their ecological benefits are generally less well documented. If an entity strongly desires non-native plants for aesthetic, cultural, or practical reasons, they should take care to make sure the species are not listed as invasive (refer to the California Invasive Plant Council). While urban forestry programs focus mainly on trees and residential home owners may primarily take interest in shrubs and understory plants, ultimately integrating many different types of native vegetation into the urban landscape will provide the largest benefit for biodiversity.

NATIVE PLANTS IN THE CITY • PHOTOS BY JOHN RUSK (CC BY 2.0), SFEI, AND MELINDA STUART (CC BY 2.0)

Plant List 1.
OAK WOODLAND, SAVANNA AND GRASSLAND

	Historical habitat types: Oak woodland, Oak savanna/grassland	
Trees	Quercus lobata	Valley oak
	Quercus agrifolia	Coast live oak
	Aesculus californica	California buckeye
	Quercus kelloggii	California black oak
	Umbellularia californica*	Bay laurel*
Shrubs/ Small trees	Corylus cornata ssp. californica**	Hazelnut**
	Heteromeles arbutifolia	Toyon
	Mimulus aurantiacus	Sticky monkeyflower
	Symphoricarpos albus	Common snowberry
	Frangula claifornica	Coffeeberry
	Lupinus albifrons	Silver lupine
	Artemesia californica	California sage
Herbaceous understory	Stipa pulchra	Purple needle grass
	Elymus glaucus	Blue wild rye
	Eschscholzia californica	California poppy
	Sisyrinchium bellum	Blue-eyed grass
	Nemophila menziesii	Baby blue eyes
	Lupinus bicolor	Minature lupine
	Clarkia purpurea	Purple clarkia
	Archillea millefolium	Yarrow
	Symphyotrichum chilense	California aster

* Do not plant within 50 ft of oak as precaution for Sudden Oak Death.

** Plant upslope of immediate riparian area.

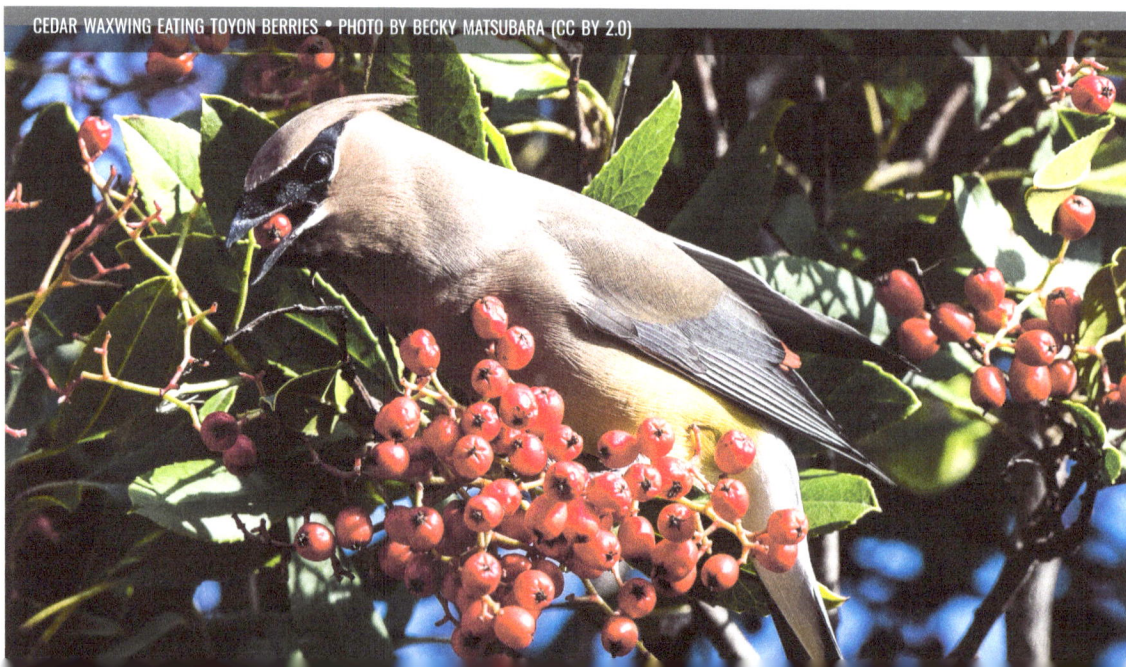

CEDAR WAXWING EATING TOYON BERRIES • PHOTO BY BECKY MATSUBARA (CC BY 2.0)

Plant List 2.
RIPARIAN COMPLEX

	A. Riparian forest Historical habitat types: *Willow thicket, Wild rose thicket, Box elder grove*		B. Riparian scrub Historical habitat types: *Sycamore alluvial woodland, riparian scrub, sparsely vegetated riparian*	
Trees	*Acer macrophylum*	Bigleaf maple		
	Acer negundo	Boxelder maple	See A.	
	Alnus rhombifolia	White alder		
	Platanus racemosa	California sycamore		
	Populus fremontii	Fremont cottonwood		
	Salix laevigata	Red willow		
	Salix lasiolepis	Arroyo willow		
Shrubs/ Small trees	*Sambucus nigra ssp. caerulea*	Blue elderberry	*Sambucus nigra ssp. caerulea*	Blue elderberry
	Cornus sericea	Red osier dogwood	*Cornus sericea*	Red osier dogwood
	Vitis californica	California grape	*Vitis californica*	California grape
	Rosa californica	California rose	*Rosa californica*	California rose
	Symphoricarpos albus	Common snowberry	*Baccharis pilularis*	Coyote bush
	Corylus cornata var. californica	Hazelnut	*Solanum umbelliferum*	Blue witch
Herbaceous understory	*Rubus ursinus*	California blackberry	*Rubus ursinus*	California blackberry
	Juncus patens	Common rush	*Juncus patens*	Common rush
	Cyperus eragrostis	Tall flatsedge	*Artemesia douglaiana*	Mugwort
	Clematis ligusticfolia	Virgin's Bower	*Baccharis salicifolia*	Mulefat
	Mimulus guttatus	Seep monkeyflower		

BLUE ELDERBERRY • PHOTO BY BRIAN (CC BY 2.0)

CALIFORNIA SYCAMORE • PHOTO BY AMY RICHEY (SFEI)

Palette 3: WET MEADOW				
	A. Alkali meadow **Historical habitat types:** *Alkali meadow*		**B. Wet meadow** **Historical habitat types:** *Wet meadow*	
Trees*	*Salix laevigata*	Red willow	*Acer macrophylum*	Bigleaf maple
	Salix lasiolepis	Arroyo willow	*Salix laevigata*	Red willow
			Acer negundo	Boxelder maple
			Alnus rhombifolia	White alder
			Platanus racemosa	California sycamore
			Populus fremontii	Fremont cottonwood
			Salix lasiolepis	Arroyo willow
Shrubs/ Small trees	*Baccharis pilularis*	Coyote bush	*Baccharis pilularis*	Coyote bush
			Morella californica	Pacific wax myrtle
			Sambucus nigra ssp. caerulea	Blue elderberry
Herbaceous understory	*Distichilis spicata*	Salt grass	*Agrostis pallens*	Diego bent grass
	*Hordeum depressum***	Alkali barley**	*Deschampsia cespi-tosa ssp. holciformis*	Pacific hairgrass
	Jaumea carnosa	Jaumea	*Elymus triticoides*	Beardless wildrye
	Frankenia salina	Alkali heath	*Festuca rubra*	Red fescue
	Limonium californicum	Marsh rosemary	*Leymus triticoides*	Creeping wild rye
	*Bacharris douglasii***	Marsh baccaris**	*Juncus patens*	Common rush
			Danthonia californica	California oatgrass
			Asclepias fascicularis	Narrowleaf milkweed

* No trees recommended near tidal marsh (e.g., 200 feet) due to concern of creating additional raptor perches (BCDC 2007).

** Nursery availability may be limited.

SELECTING THE RIGHT PLANTS

The following section highlights a few key considerations that may affect, given an established habitat type goal, the particular plants to select from or modify in a larger list (e.g., those suggested on page 26-28), how to source them, and how to adjust them through time with a changing climate.

Genetics and sourcing. Genetic diversity is an important component of the diversity of plants. The local genetic varieties of different species form distinct populations that may have unique adaptations. Where possible, it is preferable to use locally-sourced seeds to conserve local genetic diversity and adaptability, and to avoid using cultivars that may contaminate local gene pools. Work with nurseries that use best practices for avoiding the spread of diseases (e.g., Sudden Oak Death). Planting at different stages of a plant's life cycle comes with tradeoffs. Seeds (e.g. acorns) take more time to develop, while seedlings or containers may be limited in their taproot development, and in-ground container stock or transplants may be expensive or require significant maintenance. Selecting what stage to plant will require planning based on nursery availability, project timeline, and budget.

Ecology and disturbance. It is important to consider the existing vegetation and associated competitive dynamics in ecological systems for site planning. Making room for or selecting plants that can grow to their full potential is important. Excessive shade can limit the potential for a plant to produce a full crop of foliage, flowers, and fruit. In some systems like grasslands, competition can make the maintenance of native wildflowers challenging. Without regular disturbance like fire or light grazing, native woody plants or invasive species may crowd out the native grassland species. Selecting plants that will bloom at different times can provide resources for wildlife across the seasons. For flowers with nectar in the understory, consider focusing on planting many individuals of one species (rather than just diversity alone). This helps create patches to visually attract insects, whose pollination can help sustain the plant populations. For oak and riparian woodlands, planting understory plants, shrubs, and trees together can help provide a diverse and complex canopy to support wildlife (H.T. Harvey & Associates 2018). While complex habitat structure is desirable in general, alkali meadow, wet meadow and grassland were historically composed of understory plants, with little to no trees or shrubs. To create or maintain dedicated patches of these habitat types, such as in yards or parks, it may not be necessary to include trees. However, a community may desire woody vegetation such as street trees for aesthetics, shade or other values. In these cases, a list of native trees and shrubs that could tolerate similar conditions is provided.

Climate change. A changing climate may alter the suitability of both a habitat type and its consituent plant species for a given location. The Santa Clara Valley is projected to increase in temperature by at least several degrees over the next century (Fitzpatrick and Dunn 2019), and California is expected to experience more intense periods of drought (Diffenbaugh et al. 2015). Climate change may affect groundwater levels in some cases as well (e.g. Hoover et al. 2017), changing which plants are suitable in some locations.

Climate change is likely to increase the management and intervention necessary in order to conserve both habitats and species. For instance, climate shifts may increase the pace or extent of conversion of some habitat types to shrublands in the Bay Area, particularly in the absence of disturbance (Cornwell et al. 2012). In this case, managing with more regular disturbance to maintain habitat types that may otherwise convert to shrublands may be necessary if retaining these habitat types on the landscape is a goal.

While climate change will affect all biodiversity, its effects on near-term plant selection decisions is most relevant for selecting long-lived species such as trees—selecting tree species now that can withstand climate change will be important for ensuring their longevity. Some tree species, including several native California oaks, are already adapted to periodic drought and may already possess the genetic diversity within their populations necessary to ensure they can tolerate future conditions (Spotswood et al. 2017). Other tree species may have more limited heat and drought tolerances. Rather than replanting these species in their historical ranges, it may be necessary to import plants adapted to hotter, drier climates. Smaller plants with relatively short life spans (e.g. <15 years) are less likely to be significantly affected by climatic trends, and it may not be necessary to make substitutions over the short term for these species. While landscapers should prioritize locally sourced seed should be prioritized in the near term, over the longer term, it may be necessary to consider and test more drought-adapted varieties or other species from warmer climates. Further experimentation, monitoring and adaptive management will ensure long-term success of selected plants in cities.

OUR CITY FOREST COMMUNITY NURSERY, SAN JOSE CA • COURTESY OF OUR CITY FOREST

4 Practical Considerations

Many factors may affect the composition and configuration of urban greening efforts, including local policy, infrastructure, and community input. These factors can influence which species to plant, planting location, and the maintenance of projects over time. Incorporating these considerations in the design and planning phase of a project can help increase the chance of long-term success and public support for the project. This list provides some primary considerations and relevant resources to guide local urban greening projects.

COMMUNITY CONSIDERATIONS

Community participation. A critical determinant of long-term success in urban greening efforts involves working directly with communities to plan and implement projects (e.g., Carmichael and McDonough 2019, Gardner et al. 2018). Proposals to create habitat through urban greening in certain areas provides an opportunity to align use of open space with community needs. Holding open dialogue and conducting educational outreach with the community can build trust and promote the longevity, reduce the maintenance costs, and increase the value of urban greening projects in the community.

Shared use of the landscape. Urban open spaces can be designed to include both people and nature. Establishing fencing or signage around particularly ecologically valuable areas, educating park users, sustaining funds and labor for maintenance and stewardship, and working closely with the community on stewarding local sites are all necessary actions for promoting a healthy urban landscape for all.

Cultural considerations. Public preferences and values related to plant selection can both influence plant choices and affect community acceptance of projects. For example, preferences for trees with particular characteristics, such as showy flowers, can be important drivers of tree choice in residential yards. In these cases, strong existing preferences may limit the selection of native plants and trees if plants with desired characteristics do not exist. However, native plants can introduce unique value; native plants such as oaks and willows have distinctive character, medicinal and culinary uses, and cultural value, particularly among indigenous communities (Anderson 2005). The general public may not always be aware of the cultural value and services derived from native plants, though perceptions can be shifted through discussion and education. Approaches to addressing cultural preferences should be tailored to the project, and community preferences should be considered during project design to ensure local relevance and success. Planting a variety of native plants in urban areas provides an opportunity to increase the ecological value of an area while cultivating a sense of place.

Partner collaboration. Working across a variety of land uses or jurisdictions, or even for a single project, may require coordinating across several departments to implement a project, and acquiring multiple permits. Scheduling permits and aligning them with design, planning and construction timelines while working with different agencies and groups, will require close communication, collaboration and leadership from multi-stakeholder working groups.

SELECTING PLANTS

While Section III outlines general ecological guidance for planting, other implementation considerations may help guide selection of native plants, including:

Street tree diversity. Municipal urban forestry programs (such as San Jose's) often promote high diversity within their canopies, following a 10-20-30 rule for street trees—canopy composition of 10% or less of a single species, 20% or less of one genus, and 30% or less of a family. In many Santa Clara Valley cities, oak woodland species make up less than 5% of urban trees, falling well below recommended thresholds (Spotswood et al. 2017). In these cases, adding native oaks in urban greening projects could improve native tree representation while maintaining desired tree diversity.

Approved street trees. Some cities maintain lists of approved and prohibited street trees. For instance, the City of Santa Clara includes Black Oak (Quercus kelloggii) but not other native oaks on its approved street tree list. In some cases, permits can be acquired to plant trees not included on approved lists. In others, city staff may be able to provide more information on whether exceptions or additions can be made to existing lists.

Shade and light. Tall trees provide valuable shade and mitigate the urban heat island effect. The City of San Jose directs that shade trees be used wherever possible as street trees (Abeyta et al. 2013). Some cities also have shade requirements for parking lots. In such areas where existing tree canopy is sparse, large trees with dense, dark canopy such as coast live oaks may be better choices than smaller trees with more sparse canopies, such as buckeye or blue elderberry. While trees are valued for shade, they are also sometimes valued for their ability to let in some light. For example, the City of San Jose Downtown Streetscape Master Plan provides guidance to select trees that provide shade as well as filtered sunshine instead of very shady evergreen trees (SMWM and Fehr & Peers Associates, Inc. 2003). Valley oaks, sycamores, or other native deciduous trees may be appropriate candidates in these cases.

Structure. Some areas, such as the City of Milpitas, have guidance that recommends use of shrubs and groundcover without thorns or complex branching patterns (which can accumulate debris) [e.g., City of Milpitas 2000]. Opportunities to plant native plants can be arranged through simple substitutions, such as favoring snowberry over California rose in riparian areas.

SELECTING PLANTING LOCATION

The following considerations guide opportunities for where plants can be placed given land use controls.

Proximity to infrastructure. Integrating urban greening projects with infrastructure is a key component in the project design phase. For example, trees or large shrubs should be planted in locations that will allow adequate clearance from power lines, stop signs and traffic lights, underground utilities, fire hydrants, sidewalks, solar panels (see California Solar Shade Control Act), and provide sufficient visibility for vehicle and pedestrian safety. Opportunities to plant large shrubs and trees vary slightly depending on geography and jurisdiction; more specific guidance is available in materials such as the General Infrastructure guidelines from the San Jose City Department of Transportation, the San Jose Redevelopment Agency Downtown Streetscape Master Plan, City of San Jose Municipal Code, and the Milpitas Master Streetscape Plan.

Fire prevention. Some of the best locations for urban greening may lie at the urban fringe, close to large ecologically valuable open space. Urban greening opportunities here are most suitable and sustainable at least 100 feet from structures, and CalFire guidelines for defensible space should be consulted during project planning and design.

Parking requirements. Cities often have mandatory minimums for number of parking spaces, which can vary depending on zoning ordinances and facility type. These values also vary by city and may be based on building square footage or per employee or sales. Opportunities to remove and replace unnecessary or underutilized parking spaces and asphalt with vegetation will depend on these parking requirements.

OAK TREE IN SANTA CLARA VALLEY • SHIRA BEZALEL (SFEI)

MANAGING HABITAT

The following give consideration to establishing and maintaining habitat in the urban landscape:

Maintaining clearance. Requirements to clear vegetation around intersections, infrastructure (see "planting location - proximity to infrastructure") or structures for visibility or fire safety may affect the siting of plants and how they are managed. Planning ahead to site larger woody plants more to the interior of a site or away from these clearance requirements can help ensure the longevity of plants and reduce future pruning costs.

Maintaining dead vegetation. Dead limbs, trees, and leaf litter can provide valuable habitat for a variety of creatures, from birds to insects to salamanders. Setting back or fencing off areas with dead trees can benefit public safety and wildlife. Dead limbs can be cut off and placed on the ground to provide habitat. Creating large cutouts in sidewalks for street trees can help catch and retain leaves and seeds that would otherwise be removed. Reducing raking, leaf blowing, or other practices that remove leaf litter or topsoil can also improve the health of soil, vegetation and long-term ecological success of urban greening projects.

Timing and nesting season. The timing of the region's climate of wet and dry seasons, cycles of plant growth and the timing of animal migrations all affect the prioritization of management activities throughout the year. A few simple things that can be done to minimize impacts on wildlife include dimming interior lights during bird migration seasons and minimizing vegetation management during the spring and summer nesting seasons. More detail on this topic, including a more complete schedule of natural phenomena and associated management practices, can be found in SFEI-ASC 2019.

Tree ordinances. Tree ordinances are one tool to help ensure long-term persistence of trees as habitat features in an urban location. Ordinances that restrict removal or standards for pruning of long-lived trees can protect the investment of trees as ecological infrastructure in an urban landscape. In cities where ordinances are present, they can influence urban greening projects if plans include the removal of trees. Urban greening can also help cities build heritage tree populations through deliberate planting of species that will be protected when they become large.

Useful References

The following provides a set of recommended reading for more details related to implementation of urban greening projects (particularly as related to Sections 3 and 4), including general resources, plant selection and management, and municipal-specific guidelines and policy.

GENERAL RESOURCES

H.T. Harvey & Associates Ecological Consultants. 2018. *Integrating Nature into the Urban Landscape: A Design Guide.* https://www.harveyecology.com/integrating-nature-urban-landscape • *This resource provides key guidance for project design in the urban landscape, with considerations of setting goals, design parameters, landscape management, bird-safe design guidelines, and additional plant lists. It was instrumental in informing this report (particularly Chapter III, "Planting Considerations").*

H.T. Harvey & Associates Ecological Consultants. In press. *Residential Guide to Habitat Design.* • *This resource is an application of the more general design guide described above to residential areas, with specific guidance targeted to homeowners. It provides detailed guidance for at the site scale.*

San Francisco Estuary Institute. 2019. *Urban Biodiversity Framework: A science-based approach to enhancing nature in cities, with application to Silicon Valley.* **San Francisco Estuary Institute, Richmond CA.** • *This document provides comprehensive guidance at the landscape scale for how to coordinate actions across the landscape to support biodiversity. This framework is then applied with the help of spatial data to the larger Silicon Valley as an example.*

Our City Forest. http://www.ourcityforest.org/ • *A local nonprofit that provides urban forestry expertise as well as education, outreach and incentive programs. It also maintains a local plant nursery.*

EcoAtlas. 2019. San Francisco Estuary Institute. https://www.ecoatlas.org/ • *This resource provides a variety of mapping layers useful for interpreting historical ecology and natural resources information.*

PLANT SELECTION AND MANAGEMENT

Calscape. California Native Plant Society. http://calscape.org/ • *This website provides detailed information on plant distribution records as well as habitat conditions for plant species suitability and tolerances. This resource is particularly helpful for verifying that a particular plant species may be suitable for a given local site and plant community.*

Popper, H. 2012. California Native Gardening—a month-by-month guide. University of California Press. • *This book provides approachable guidance for native plant gardening, with an emphasis and organization centered around the blooming schedule of native plants.*

Bauer N. 2012. *The California Wildlife Habitat Garden: How to Attract Bees, Butterflies, Birds and Other Animals.* **University of California Press.** • *This book provides guidance on associations between native plants and wildlife species, for practitioners interested in attracting particular wildlife species.*

[BCDC] San Francisco Bay Conservation and Development Commission. 2007. *Shoreline Plants: A Landscape Guide for the San Francisco Bay.* http://www.bcdc.ca.gov/planning/SPLG.pdf • *Information on potential candidates for plant species, particularly in relation to planting sites close to the Bay.*

Costello et al. 2013. *Oaks in the Urban Landscape: Selection, Care, Preservation.* University of California Agriculture and Natural Resources Publication 3518. Richmond, CA. • *This document provides resources for specifically related to oak planting management in the urban landscape.*

MUNICIPAL-SPECIFIC GUIDANCE

Abeyta D, Lanham E, Hansen R, Mize R, Our City Forest. 2013. *San Jose Tree Policy Manual & Recommended Best Practices.* City of San Jose. http://www.sanjoseca.gov/DocumentCenter/View/21896

SMWM, Fehr & Peers Associates, Inc. 2003. *San Jose Redevelopment Agency Downtown Streetscape Master Plan.* The Redevelopment Agency of the City of San Jose. http://www.sjredevelopment.org/PublicationsPlans/streetscape.pdf

City Tree Specifications. 2007. City of Morgan Hill Public Works Department. http://www.morgan-hill.ca.gov/DocumentCenter/View/16525

City Trees. 2019. The City of Santa Clara Public Works Department. Accessed 4 April 2019. http://santaclaraca.gov/government/departments/public-works/maintenance-operations/city-trees

City of Milpitas Streetscape Master Plan. 2000. City of Milpitas. http://www.ci.milpitas.ca.gov/_pdfs/plan_mp_streetscape.pdf

OTHER POLICIES/GUIDANCE

Fire Safety Education. 2019. CalFire. http://www.fire.ca.gov/communications/communications_firesafety_100feet • *This website provides information for planning and managing projects for defensible space at the urban-wildland interface.*

Solar Rights: Access to the Sun for Solar Systems. 2019. Go Solar California. State of California, California Energy Commission & California Public Utilities Commission. https://www.gosolarcalifornia.ca.gov/solar_basics/rights.php • *This resource provides information on The Solar Rights Act and the Solar Shade Control Act, legislation that may affect tree planting and maintenance programs and activities.*

Bernhardt, E, Swiecki, TJ. 1991. *Guidelines for developing and evaluating tree ordinances.* Prepared for: Urban Forestry Program, California Department of Forestry and Fire Protection, Sacramento, CA. http://phytosphere.com/treeord • *This site and publication provide information on establishing tree protection ordinances.*

Citations

Abeyta D, Lanham E, Hansen R, Mize R, Our City Forest. 2013. San Jose Tree Policy Manual & Recommended Best Practices.

Anderson M. 2005. *Tending the wild: Native American knowledge and management of California's natural resources*. University of California Press.

Aronson MF, Lepczyk CA, Evans KL, Goddard MA, Lerman SB, MacIvor JS, Nilon CH, Vargo T. 2017. Biodiversity in the city: key challenges for urban green space management. *Frontiers in Ecology and the Environment* 15(4):189–196.

Barbour M, Keeler-Wolf T, Schoenherr A. 2007. *Terrestrial Vegetation of California*, (Third edition). Berkeley, CA: University of California Press.

Bauer N. 2012. *The California Wildlife Habitat Garden: How to Attract Bees, Butterflies, Birds and Other Animals*. University of California Press.

[BCDC] San Francisco Bay Conservation and Development Commission. 2007. Shoreline Plants: A Landscape Guide for the San Francisco Bay. :50.

Beller E, Robinson A, Grossinger R, Grenier JL. 2015. Landscape Resilience Framework: Operationalizing ecological resilience at the landscape scale.

Beller E, Salomon M, Grossinger R. 2010. Historical Vegetation and Drainage Patterns of Western Santa Clara Valley.

Beller EE, Spotswood EN, Robinson AH, Anderson MG, Higgs ES, Hobbs RJ, Suding KN, Zavaleta ES, Grenier JL, Grossinger RM. 2019. Building Ecological Resilience in Highly Modified Landscapes. *BioScience* 69(1):80–92.

Beninde J, Veith M, Hochkirch A. 2015. Biodiversity in cities needs space: a meta-analysis of factors determining intra-urban biodiversity variation. *Ecology Letters* 18(6):581–592.

Carmichael CE, McDonough MH. 2019. Community Stories: Explaining Resistance to Street Tree-Planting Programs in Detroit, Michigan, USA. *Society & Natural Resources* :1–18.

City of Milpitas. 2000. City of Milpitas Streetscape Master Plan.

Cooper Marcus C, Barnes M. 1995. *Gardens in the healthcare facilities: uses, therapeutic benefits, and design recommendations*. Berkeley, Calif.: University of California.

Cornwell WK, Stuart SA, Ramirez A, Dolanc CR, Thorne JH, Ackerly DD. 2012. Climate change impacts on california vegetation: physiology, life history, and ecosystem change. 90.

County of Santa Clara, City of San Jose, City of Gilroy, Santa Clara Valley Water District, Santa Clara Valley Transportation Authority. 2012. Final Santa Clara Valley Habitat Plan. Santa Clara County, California.

Dettinger MD, Ralph FM, Das T, Neiman PJ, Cayan DR. 2011. Atmospheric Rivers, Floods and the Water Resources of California. *Water* 3(2):445–478.

Diffenbaugh NS, Swain DL, Touma D. 2015. Anthropogenic warming has increased drought risk in California. *Proceedings of the National Academy of Sciences* 112(13):3931–3936.

eBird. 2019. eBird: An online database of bird distribution and abundance [web application]. eBird, Ithaca, New York. Available: http://www.ebird.org.

Escobedo FJ, Kroeger T, Wagner JE. 2011. Urban forests and pollution mitigation: analyzing ecosystem services and disservices. *Environmental pollution* 159(8):2078–2087.

Fitzpatrick MC, Dunn RR. 2019. Contemporary climatic analogs for 540 North American urban areas in the late 21st century. *Nature Communications* 10(1).

Forman RTT. 2016. Urban ecology principles: are urban ecology and natural area ecology really different? *Landscape Ecology* 31(8):1653–1662.

Galloway D, Jones D, Ingebritsen S. 1999. Land Subsidence in the United States. *US Geological Survey Survey* 1182.

Gardner J, Marpillero-Colomina A, Begault L. 2018. Inclusive Healthy Places: A Guide to Inclusion & Health in Public Space: Learning Globally to Transform Locally.

Goddard MA, Dougill AJ, Benton TG. 2010. Scaling up from gardens: biodiversity conservation in urban environments. *Trends in Ecology & Evolution* 25(2):90–98.

Griffin J, Critchfield W. 1972. The distribution of forest trees in California, California Pacific SW. Forest and Range Experiment Station. *USDA Forest Service Research Paper PSW* 82(114).

Grossinger R, Askevold R, Striplen C, Brewster E, Pearce S, Cayce K, McKee L, Collins J. 2006. Coyote Creek Watershed Historical Ecology Study: Historical Conditions and Landscape Change in the Eastern Santa Clara County, California.

Grossinger R, Beller E, Salomon M, Whipple A, Askevold R, Striplen C, Brewster E, Leidy R. 2008. South Santa Clara Valley Historical Study, including Soap Lake, the Upper Pajaro River, and Llagas, Uvas-Carnadero, and Pacheco Creeks.

Hoover DJ, Odigie KO, Swarzenski PW, Barnard P. 2017. Sea-level rise and coastal groundwater inundation and shoaling at select sites in California, USA. *Journal of Hydrology: Regional Studies* 11:234–249.

Karlinsky S, Szambelan S, Wang K. 2017. Room for More: SPUR's Housing Agenda for San Jose. San Francisco Bay Area Planning and Urban Research Association (SPUR).

Kasten K, Stenoien C, Caldwell W, Oberhauser KS. 2016. Can roadside habitat lead monarchs on a route to recovery? *Journal of Insect Conservation* 20(6):1047–1057.

Li W, Saphores J-DM, Gillespie TW. 2015. A comparison of the economic benefits of urban green spaces estimated with NDVI and with high-resolution land cover data. *Landscape and Urban Planning* 133:105–117.

Morey, S. 2014. Arboreal Salamander, Aneides lugubris. California Wildlife Habitat Relationships System. California Department of Fish and Wildlife, California Interagency Wildlife Task Group. https://www.wildlife.ca.gov/Data/CWHR/Life-History-and-Range

Payne L, Orsega-Smith B, Godbey G, Roy M. 1998. Local parks and the health of older adults. *Parks & Recreation* 33(10).

Pearson DG, Craig T. 2014. The great outdoors? Exploring the mental health benefits of natural environments. *Frontiers in Psychology* 5.

Riparian Corridor Protection and Bird-Safe Design, Policy Number 6-34, City of San Jose California Council Policy. 2016. . City of San Jose, California.

Robinson A, Beller E, Grossinger R, Grenier L. 2015. Vision for a Resilient Silicon Valley Landscape. San Francisco Estuary Institute, SFEI Contribution No. 753.

Romanow K, Kantak A, Hughey R, Walesh K, Ortbal J, Office of the Mayor, City of San Jose Staff, PwC, WBCSD Sustainable Lifestyle. 2018. Climate Smart San Jose: A People-Centered Plan for a Low-Carbon City.

San Jose Code of Ordinances, Title 20 - Zoning. 2019.

Sandifer PA, Sutton-Grier AE, Ward BP. 2015. Exploring connections among nature, biodiversity, ecosystem services, and human health and well-being: Opportunities to enhance health and biodiversity conservation. *Ecosystem Services* 12:1–15.

Sawyer Jr. J, Keeler-Wolf T, Evens J. 2009. *A Manual of California Vegetation*, (Second edition). California Native Plant Society.

SMWM, Fehr & Peers Associates, Inc. 2003. San Jose Downtown Streetscape Master Plan. The Redevelopment Agency of the City of San Jose.

[SFEI-ASC] San Francisco Estuary Institute-Aquatic Science Center. 2019. Urban Biodiversity Framework: A science-based approach to enhancing nature in cities, with application to Silicon Valley. San Francisco Estuary Institute, Richmond, CA.

Spotswood E, Grossinger R, Hagerty S, Beller E, Grenier JL, Askevold R. 2017. Re-Oaking Silicon Valley: Building Vibrant Cities with Nature.

Stanford B, Grossinger R, Beagle J, Askevold R, Leidy R, Beller E, Salomon M, Striplen C, Whipple A. 2013. Alameda Creek Watershed Historical Ecology Study. San Francisco Estuary Institute, Richmond, CA.

Threlfall CG, Mata L, Mackie JA, Hahs AK, Stork NE, Williams NSG, Livesley SJ. 2017. Increasing biodiversity in urban green spaces through simple vegetation interventions. *Journal of Applied Ecology* 54(6):1874–1883.

Zeiner D, Laudenslayer, Jr. W, Mayer K, White M. 1988. California's Wildlife. Accessed from https://www.wildlife.ca.gov/Data/CWHR/Life-History-and-Range, 2019-3-1.

Appendix

PLANT METHODS

The plant lists are comprised of plants generally thought to be native to the Santa Clara Valley, and where possible to be determined, the Valley floor. These lists were synthesized from a variety of resources. *Terrestrial Vegetation of California* (Barbour et al. 2007) and Holland Classifications from the *Manual of California Vegetation* (Sawyer Jr. et al. 2009) were used to build lists of general plant community associates. The *California Wildlife Habitat Garden* (Bauer 2012) and BCDC Shoreline Plants (BCDC 2007) were used for additional inspiration. Historical ecology data (Grossinger et al. 2006, Grossinger et al. 2008, Beller et al. 2010) including General Land Office surveys, Griffin and Critchfield 1972 and CalFlora were used to assess historical local native status where information was readily available, and CNPS' Calscape tool interpolated ranges were used to confirm contemporary distribution, particularly for understory plants. Calscape was also used to verify and assess plant suitability for habitat type, moisture requirements, salinity tolerance and commercial nursery availability. Consideration was also given to plant aesthetics. Finally, local expert opinion, provided by the technical advisory committee for this report, helped revise and polish these lists.

Table A1. Crosswalk of historical habitat types and corresponding plant lists

Historical habitat type	Corresponding plant list
Oak Savanna / Grassland, Oak Woodland	List 1: Oak woodland/savanna and grassland
Box Elder grove, Wild Rose Thicket, Willow thicket	List 2a: Riparian forest
Sycamore alluvial woodland, riparian scrub and sparsely vegetated riparian	List 2a: Riparian scrub
Alkali Meadow	List 3a: Alkali meadow
Wet Meadow	List 3b: Wet meadow
Deep Bay, Shallow Bay/Channel, Perennial Pond, Seasonal Lake/Pond, Chaparral, Valley Freshwater Marsh	Not listed; none associated. Not all historical habitat types were linked to plant lists for this report, which focused more on historically prevalent terrestrial habitat types. Habitats of low historical acreage or aquatic habitats were not linked to plant lists.

www.ingramcontent.com/pod-product-compliance
Lightning Source LLC
Chambersburg PA
CBHW050917210326
41597CB00003B/131